# DARWIN'S TREE OF DEATH

## THE CONSEQUENCE
## OF
## 'THE SURVIVAL OF THE FITTEST'

# Darwin's Tree of Death

## The Consequence of 'The Survival of the Fittest'

A tree is recognised by its fruit (Matthew 12:33).

### David H.J.Gay

**BRACHUS**

BRACHUS 2018
davidhjgay@googlemail.com

Scripture quotations come from a variety of versions

All my work may now be found on davidhjgay.com

# Contents

Preamble ..................................................................... 9
Introduction ............................................................... 11
Extinction: The Track Record ................................. 15
Summary .................................................................. 25
Three Questions Asked and Answered ..................... 29
A Positive Note to End On ....................................... 35

# *Preamble*

According to Woody Allen, writing in 1979, mankind faces a dire choice:

> More than [at] any other time in history, mankind faces a crossroads. One path leads to despair and utter hopelessness. The other, to total extinction. Let us pray we have the wisdom to choose correctly.[1]

Forty years later, have things have got any better?

---

[1] Woody Allen: 'My Speech to the Graduates', *The New York Times*, August 10th, 1979.

# Introduction

My tract – 'So... You Believe in Evolution?' – has not met with universal approval. Surprise, surprise! Many advocates of evolution do not like the idea that their system, their philosophy, their religion – make no mistake, that is what evolution is, a religion, a system of belief – inevitably leads to consequences they simply dare not face. Some evolutionists, I acknowledge, have been willing to own the consequences of their system, but many deny them, or otherwise bury their head in the sand, preferring not to think of the outcome of their system.[1]

But it will not do!

First, let me say just a little more about evolution as a religion. Though many evolutionists may try to deny this and talk loftily about a science, they are clearly mistaken; science needs a human being to observe and measure events, changes, phenomena.[2] Evolution, by its very definition, pre-dates man. So let's make no bones about it (no pun intended): evolution is not a science; it is a belief system, a philosophy, a religion.

In short, to believe the universe – man, in particular – is the product of evolution is no different to believing that the universe was created by Almighty God; both are belief systems. Evolutionists are believers; they believe in 'natural selection'.

---

[1] While the response to my tract played its part, the immediate catalyst for this article has been Andrew Marr's 'Darwin's Dangerous Idea'.

[2] As the *Oxford Dictionary's* website has it: 'Science is the intellectual and practical activity encompassing the systematic study of the structure and behaviour of the physical and natural world through observation and experiment' – study and observation by man, of course.

## Introduction

Now, as with all thought systems, all religions, the theory of evolution has consequences. Evolutionists like to talk about the tree of life; the tree of *death* is more like it! And inevitable death, at that! Indeed, extinction is inevitable; indeed, it is an essential factor in the process of evolution – one which needs no apology; rather, it should be gloried in. This is what evolution demands and what it leads to. Many of its advocates may not like it, but it is the stubborn truth, and they need to face up to it!

As I imply, some do.

Take this abstract at the head of G.Beer's 'Darwin and the uses of extinction':

> We currently view extinction with dismay and even horror, but Darwin saw extinction as ordinary and as necessary to evolutionary change. Still, the degree to which extinction is fundamental to his theory is rarely discussed. This essay ['Darwin and the uses of extinction'] examines Darwin's linking of the idea of 'improvement' with that of natural selection and tracks a cluster of reasons for our changed valuation of extinction now. Those reasons demonstrate how scientific information and ideological preferences have reshaped the concept. [This] essay challenges the reader to assess some current assumptions about extinction, and concludes by considering the shift in Darwin's own understanding from the *Origin* to the late *Autobiography*.

So, let's go back to the beginning – to Charles Darwin himself. In his *On the Origin of Species* (1859), Darwin stated in black and white:

> Extinction and natural selection... go hand in hand.

> Each new variety, and ultimately each new species, is produced and maintained by having some advantage over those with which it comes into competition; and the consequent extinction of the less-favoured forms almost inevitably follows.

> The extinction of species and of whole groups of species, which has played so conspicuous a part in the history of the organic world, almost inevitably follows on the principle of

## Introduction

natural selection; for old forms will be supplanted by new and improved forms.

Darwin, summing up his book, listed what he described as the fundamental components or laws of the evolutionary process: reproduction, inheritance, variability, the struggle for life, and natural selection, with its 'consequences', one of which was the extinction of less-improved forms.

Darwin, in his *The Descent of Man* (1871), declared:

> At some future period, not very distant as measured by centuries, the civilised races of man will almost certainly exterminate and replace throughout the world the savage races. At the same time the anthropomorphous apes... will no doubt be exterminated. The break will then be rendered wider, for it will intervene between man in a more civilised state, as we may hope, than the Caucasian, and some ape as low as a baboon, instead of as at present between the negro or Australian and the gorilla.

Thus it is patent. Right from Darwin himself, we have been left in no doubt: if evolution is right, extinction is inevitable; more, it is commendable, it is essential, and needs no apology. And mankind is as much bound up in inevitable extinction as are insects, microbes, viruses, and the like.

This is not to say that Darwin would have approved of all that has been done in the name of his teaching. But, the sluice having been opened, nobody should be surprised at the ensuing flood.

Everybody knows that evolution means 'the survival of the fittest'. Darwin did not coin the phrase – that was the work of Herbert Spencer (1820-1903):

> This survival of the fittest, which I have here sought to express in mechanical terms, is that which Mr Darwin has

## Introduction

called 'natural selection', or the preservation of favoured races in the struggle for life.[3]

Nevertheless, though Darwin did not coin the phrase, he soon latched on to it.[4] 'The survival of the fittest' became the evolutionist's mantra. And, as night follows day, 'the survival of the fittest' inevitably means 'the extinction of the un-fittest'.

Let me sketch some of the main points in the history of extinction on the basis of the survival of the fittest.

---

[3] See the Darwin Correspondence Project for July 1866; Herbert Spencer: *The Principles of Biology*, University Press of the Pacific, 2002, p444; Wikipedia.

[4] See Charles Darwin: *The Variation of Animals and Plants under Domestication* (1868); *On the Origin of Species* (fifth edition, 1869).

# Extinction: *The Track Record*

Take Karl Marx (1818-1883), who argued for struggle in the economic field just as Darwin had argued for struggle in the natural world. As Augosto Zimmermann said of Marx:

> Marx believed that the existence of God had been disproved by the inexorable forces of science, reason and progress. As such, Darwinism became an important element of Marxist theory. As his close friend and co-writer Friedrich Engels pointed out, 'Just as Darwin discovered the law of evolution in organic nature, so Marx discovered the law of evolution in human history'. In a personal letter to him, Marx actually reveals that Darwin's *Origin of Species* was indeed very important, as it had provided him 'with the basis in natural science for the class struggle in history'. As a sign of gratitude, Marx sent Darwin the second German edition of *Das Kapital*. On the title page he inscribed, 'Mr Charles Darwin/On the part of his sincere admirer/[signed] Karl Marx, London 16 June 1873'.[1]

Take Friedrich Nietzsche (1844-1900). As R.J.Hollingdale put it:

> Darwin had shown that the higher animals and man could have evolved in just the way they did entirely by fortuitous variations in individuals. Natural selection was for Nietzsche essentially evolution freed from every metaphysical implication.[2]

In his *The Gay Science* (1882) and again in his *Thus Spoke Zarathustra* (1883-1891), Nietzsche baldly declared that 'God is dead'. He did not shirk the inevitable consequence. He was convinced that taking pity on another is not good; showing pity – Christian pity, in particular – is deplorable, a fatal weakness, a mark of degeneration, a hallmark of a slave

---

[1] Augosto Zimmermann: 'Marxism, law and evolution: Marxist law in both theory and practice'.
[2] R.J.Hollingdale: *Nietzsche: The Man and his Philosophy* (1965, revised 1999).

mentality. The evolutionary principle demands self-interest above all:

> 'Nihilism' becomes an important concept for the late Nietzsche. He anticipated a crisis that will follow the gradual decline of Christianity... That all moral values such as altruism, humility, compassion are null and void is the fallout of this crisis.[3]

Take his *Twilight of the Idols* (1888):

> Self-interest is worth as much as the person who has it: it can be worth a great deal, and it can be unworthy and contemptible. Every individual may be scrutinised to see whether he represents the ascending or the descending line of life. Having made that decision, one has a canon for the worth of his self-interest. If he represents the ascending line, then his worth is indeed extraordinary – and for the sake of life as a whole, which takes a step farther through him, the care for his preservation and for the creation of the best conditions for him may even be extreme.

Take his *The Antichrist* (1895):

> Christianity is called the religion of pity. Pity stands opposed to the tonic emotions which heighten our vitality! It has a depressing effect. We are deprived of strength when we feel pity. That loss of strength which suffering as such inflicts on life is still further increased and multiplied by pity. Pity makes suffering contagious. Under certain circumstances, it may engender a total loss of life and vitality out of all proportion to the magnitude of the cause.

> [Pity is to be deplored because] pity preserves things that are ripe for decline; it defends things that have been disowned and condemned by life, and it gives a depressive and questionable character to life itself by keeping alive an abundance of failures of every type. People have dared to call pity a virtue... and as if this were not enough, it has been made the virtue, the basis and source of all virtues.

> Pity is the practice of nihilism. To repeat: this depressive and contagious instinct crosses those instincts which aim at

---

[3] 'Marx, Nietzsche and Freud' (Open University, 2016).

the preservation of life and at the enhancement of its value. It multiplies misery and conserves all that is miserable, and is thus a prime instrument of the advancement of decadence: pity persuades men to nothingness!

In short, according to Nietzsche, working on the principles of evolution, to show pity is odious. Because of the benefits of 'the survival of the fittest', we should glory in 'the extinction of the weakest'.

Francis Galton (1822-1911). Here is the abstract of an article (in French) by D.Aubert-Marston:

> Not only was Sir Francis Galton a famous geographer and statistician, he also invented 'eugenics' in 1883. Eugenics – defined as the science of improving racial stock – was developed from a new heredity theory, conceived by Galton himself, and from the evolution theory of Charles Darwin, transposed to human society by Herbert Spencer. Galton's eugenics was a program to artificially produce a better human race through regulating marriage and thus procreation. Galton put particular emphasis on 'positive eugenics', aimed at encouraging the physically and mentally superior members of the population to choose partners with similar traits. In 1904, he presented his ideas in front of a vast audience of physicians and scientists in London. His widely-publicised lecture served as the starting point for the development of eugenics groups in Europe and the United States during the first half of the 20th century.[4]

Take H.G.Wells (1866-1946). As Jerry Bergman said:

> After being exposed to Darwinism in school, H.G.Wells converted from devout Christian to devout Darwinist and spent the rest of his life proselytising for Darwin and eugenics.[5] Wells advocated a level of eugenics that was even more extreme than Hitler's [when his time came]. The weak should be killed by the strong, having 'no pity and less benevolence'. The diseased, deformed and insane, together with 'those swarms of blacks, and brown, and dirty-white, and yellow people... will have to go' in order to

---

[4] 'Sir Francis Galton: The Father of Eugenics'.
[5] Exactly! Darwinism *is* a religion, as I claimed.

create a scientific utopia. He envisioned a time when all crime would be punished by death because 'People who cannot live happily and freely in the world without spoiling the lives of others are better out of it'. He was hailed as an 'apostle of optimism', but died an 'infinitely frustrated' and broken man, concluding that 'mankind was ultimately doomed and that its prospect is not salvation, but extinction'. Despite all the hopes in science, the end must be 'darkness still'. Wells' life abundantly illustrates the bankruptcy of consistently applied Darwinism.[6]

Winston Spencer Churchill (1874-1965) bought heavily into eugenics. As Martin Gilbert said:

> He [Churchill] wrote to the Prime Minister, H.H.Asquith, in December 1910, about the 'multiplication of the unfit' that constituted 'a very terrible danger to the race'. Until the public accepted the need for sterilization, Churchill argued, the 'feeble-minded' would have to be kept in custodial care, segregated both from the world and the opposite sex.
> In his letter, Churchill told Asquith: 'The unnatural and increasingly rapid growth of the feeble-minded and insane classes, coupled as it is with a steady restriction among all the thrifty, energetic and superior stocks, constitutes a national and race danger which it is impossible to exaggerate. I am convinced that the multiplication of the feeble-minded, which is proceeding now at an artificial rate, unchecked by any of the old restraints of nature, and actually fostered by civilised conditions, is a terrible danger to the race'. Concerned by the high cost of forced segregation, Churchill preferred compulsory sterilization to confinement, describing sterilization as a 'simple surgical operation so the inferior could be permitted freely in the world without causing much inconvenience to others'.
> In February 1911, Churchill spoke in the House of Commons about the need to introduce compulsory labour camps for 'mental defectives'. As for 'tramps and wastrels', he said, 'there ought to be proper labour colonies where they could be sent for considerable periods and made to realise their duty to the State'. Convicted criminals would be sent to these labour colonies if they were judged 'feeble-minded' on medical grounds. It was estimated that some

---

[6] 'H.G.Wells: Darwin's Disciple and Eugenicist Extraordinaire'.

## Extinction: The Track Record

20,000 convicted criminals would be included in this plan. To his Home Office advisers, with whom he was then drafting what would later become the Mental Deficiency Bill, Churchill proposed that anyone who was convicted of any second criminal offence could, on the direction of the Home Secretary, be officially declared criminally 'feeble-minded', and made to undergo a medical enquiry. If the enquiry endorsed the declaration of 'feeble-mindedness', the person could then be detained in a labour colony for as long as was considered a suitable period.[7]

The Eugenics Record Office (ERO) in Cold Spring Harbor, New York, established by the Carnegie Institution of Washington's Station for Experimental Evolution, was founded by Charles Benedict Davenport (1866-1944), and directed by Harry H. Laughlin (1880-1943). It was a research institute which, from 1910-1938, gathered biological and social information about the American population. Its mission was to collect substantial information on the ancestry of the American population, to produce propaganda that was made to fuel the eugenics movement, and to promote the idea of race-betterment.

As the ERO website makes clear:

> Eugenics was and continues to be a controversial issue due to the pressure radical eugenicists put on the government to pass legislation that would restrict the liberties of the people who had traits that could be considered undesirable. Specifically, the ERO dedicated its resources to the restriction of immigrants and the forced sterilization of individuals with undesirable characteristics. They promoted their ideas through the distribution of propaganda that came in the form of images and information packets.

All was not plain sailing however:

---

[7] Martin Gilbert: 'Churchill and Eugenics'. The Mental Deficiency Act became law in 1913; only three MPs voted against it – one being Josiah Wedgewood who declared: 'It is a spirit of the Horrible Eugenic Society which is setting out to breed up the working class as though they were cattle' (Wikipedia).

## Extinction: The Track Record

Something else that caused tension within and surrounding the ERO was Harry H.Laughlin's radical policy suggestions. He was known for presenting fraudulent evidence to support policies of forced sterilization and was known for dogmatism.

And many could not live with the stark consequences of their belief:

> Furthermore, the rise of Nazism in the 1930s and their use of and belief in eugenics led to large criticism and the ultimate closing of the ERO and their practices.

Nevertheless:

> As the wave of new adherents to the eugenic cause increased, the movement took on an almost religious or spiritual character. Charles Davenport, Director of the Eugenics Record Office, came to realise the need for a canon or dogma for the movement and developed the *Eugenics Creed*. The creed underwent some minor modifications over time, but it remained the central doctrine of the eugenics crusade.[8]

Enter George Price (1922-1975). This is what Michael Regnier wrote about Price:

> It was altruistic ants that posed a particular problem for Charles Darwin. Natural selection is often described as 'survival of the fittest', where fitness means how successful an individual is at reproducing. If one individual has a trait that gives them a fitness advantage, they will tend to have more offspring than the others; because the advantage is likely to be passed on to their offspring, that trait will then spread through the population. A fundamental part of this idea is that individuals are competing for the resources they need to reproduce, and fitness includes anything that helps an individual reproduce more than the competition.
> But as Darwin observed, ants and other social insects are not in competition. They are cooperative, to the extent that worker ants are sterile and so have literally zero fitness. They ought to be extinct, yet there they are in every generation sacrificing their own reproductive ambitions to

---

[8] 'Charles Davenport and the Eugenics Record Office'.

## Extinction: The Track Record

serve the fertile queen and her drones. Darwin suggested that competition between groups of ants – queen, drones, and workers together – might be driving natural selection in this case. What was good for a nest competing against other nests would then outweigh what was good for any individual ant.

Group selection, as this idea was known, was not a very good solution, though. It didn't explain how the cooperative behaviour evolved in the first place. The first altruistic ant would have been at such a huge disadvantage compared to the rest of its group that it would never have got the chance to breed more altruistic ants. The same was true of humans – natural selection was intrinsically stacked against any altruistic individual surviving long enough to pass on their altruism.

This left a rather embarrassing paradox: The evolution of altruism was impossible, yet clearly altruism had evolved. If the biologists couldn't resolve this, would they have to throw out the whole idea of natural selection?

Luckily, a young man called Bill Hamilton spared biology's blushes with a slightly different solution in 1964. He proposed that altruism could have evolved within family groups – yes, an individual altruist would seem to be at a disadvantage, but that was not the whole picture because other individuals who shared the same genes associated with altruism would all influence each other's 'inclusive fitness'.

Discussions of human altruism are often framed in terms of someone drowning in a pond. Do you put your own life at risk to try and save them? If you do, that's altruism. Hamilton's idea, which became known as 'kin selection', acknowledged that compared to a selfish person who never got their feet wet, someone who went around jumping into ponds to save drowning people would be at a greater risk of dying before they managed to reproduce and pass their altruistic genes on to their children. However, if they happened to save a relative who shared the same genes, our altruist would have indirectly helped to get those genes passed on to the next generation after all. If the total benefit derived from having altruistic genes in the family, so to speak, was greater than the cost, then the evolution of altruism was no longer paradoxical.

## Extinction: The Track Record

When George Price stumbled across Hamilton's work in the Senate House Library in 1968, he was shocked. He was forced to confront the relationship between morality and family, the biological imperative he should have felt to sacrifice his selfish ambitions in favour of supporting his kin. He immediately set to work to challenge, even disprove Hamilton's theory. But he could only confirm it. Along the way, he derived his equation of natural selection, which helped to prove that altruism was not selfless and moral, but rather selfish and genetic.[9]

What of the aforementioned W.D.Hamilton (1936-2000)? Just this:

> The ultimate criterion which determines whether [a gene] $G$ will spread is not whether the behaviour is to the benefit of the behaver, but whether it is to the benefit of the gene $G$... With altruism this will happen only if the affected individual is a relative of the altruist, therefore having an increased chance of carrying the gene.[10]

Natalie Angier expressed it thus:

> Dr Hamilton... recast the concept of fitness, that is, an individual's success in reproducing, to incorporate the survival and reproductive success of the creature's close relatives – hence the term 'inclusive fitness'. In so doing, he merged Darwin's focus on individual animals competing for the privilege of siring the next generation with Mendel's studies of how distinct genetic traits are transmitted over time. The idea can be roughly understood by one biologist's remark in a pub that he would 'gladly die for two brothers, four cousins or eight second cousins' [but not for just one brother or for anybody less 'fit' than himself], each of them carrying the requisite percentage of the individual's genes to compensate for the mortal deed.[11]

---

[9] Michael Regnier: 'How Discovering an Equation for Altruism Cost George Price Everything', June 16th 2017.

[10] W.D.Hamilton: 'The Evolution of Altruistic Behaviour' in *The American Naturalist*, Sept. – Oct. 1963, The University of Chicago Press, for the American Society of Naturalists, pp354-355.

[11] Natalie Angier: 'William Hamilton, 63, Dies; An Evolutionary Biologist', *The New York Times*, March 10th, 2000.

## Extinction: The Track Record

George C.Williams (1926-2010), with his *Adaptation and Natural Selection* (1966), popularised Hamilton's theory.

Richard Dawkins (born 1941), with his *The Selfish Gene* (1976), built on Williams' work. Do not miss the title.

Think of abortion-on-demand, which is now being carried out on an industrial scale:

> In total, there were 194,668 abortions notified as taking place in England and Wales in 2017.[12]
>
> Termination is being fashioned into a tool of the new eugenics, under which we not only have the right to have a baby – regardless of our life circumstance and the type of assistance required ('gestational carriers') to obtain our entitlement – but also a right *to the baby we want*.[13]

Where does this come from? Take a look at this:

> These articles (on the 'Abortions and Eugenics' website) are concerned with two separate but related things: the eugenics roots of the pro-abortion movement, and the connection between modern legalised abortion and the spectre of eugenics... Margaret Sanger was a pioneer of legalised abortion and the founder of Planned Parenthood, the largest provider of abortions in the US. Sanger was also a well-known eugenicist whose writings make clear the historic connection between abortion-on-demand and eugenic thinking.[14]

In short, this kind of thinking, this kind of talk – this kind of behaviour towards the unborn child – is only possible where Darwin's 'Tree of Death' is thought to explain – and, above all, to justify – everything.

Whatever the metaphysics, the message is clear. Evolution can only be essentially self-centred; violence is rooted in the

---

[12] Official Government figures, Department of Health & Social Care, 'National Statistics' (gov.uk).
[13] Wesley J.Smith: 'Abortion and the New Eugenics', emphasis original.
[14] See 'Abortions and Eugenics'.

genes; violence, struggle for survival and reproduction of the fittest, is inevitable and unstoppable.

Lee Alan Dugatkin summarised it thus:

> One of the central dogmas of modern behavioural ecology is that blood kinship plays a critical role in understanding the evolution of social behaviour, particularly of costly social behaviour such as altruism and cooperation. But it was not always so, and what I would like to do in this *Perspectives* is provide some historical context that led up to William Hamilton's seminal work developing inclusive fitness theory. The story begins, not surprisingly, with Charles Darwin.[15]

---

[15] Lee Alan Dugatkin: 'Inclusive Fitness Theory from Darwin to Hamilton', *Genetics*, 2007.

# *Summary*

Once you accept Darwin's postulate, morality, the sense of right and wrong, the sense of fairness, goes out of the window. There is no accountability: man does not have to answer to a non-existent Almighty. You are left with selfishness and the perpetual struggle for the survival of the fittest. But this does not provide the happiness it seems to offer![1] Man merely has to struggle for survival, come what may – the 'best' man will always win, and the weakest will go to the wall! Nor should we have any regrets about it; indeed, we should be glad of it, and rejoice in the 'improvement' of the race.

Although it is a blatant – and extreme – example of what I am talking about, never forget the Nazi party under Adolf Hitler (1889-1945)![2] Never forget its blatant racism, its gross breeding-programme designed to produce a superior Germanic Aryan race, and its policy of forced sterilization, genocide and industrial holocaust to eliminate the 'unfit' –

---

[1] Jean-Paul Sartre (1905-1980): 'Everything is permissible if God does not exist, and as a result man is forlorn, because neither within him nor without does he find anything to cling to. He can't start making excuses for himself' (Jean-Paul Sartre: *Existentialism and Humanism*, 1945). But as he also said: 'That God does not exist, I cannot deny; that my whole being cries out for God I cannot forget'.

[2] Although Hitler needed no encouragement in his fervour for the application of Darwinism, Erich Ludendorff (1865-1937) did nothing to dampen it. Ludendorff – Quartermaster of the German Army in World War I – was a social Darwinist who looked on peace as just an abnormal interlude in the normality of war, war being 'the foundation of human society'. An anti-Semite, Ludendorff detested Christianity; he played a not-insignificant part in Hitler's rise to power. 'War, the belief in violence and the right of the stronger, were not the corruptions of Nazism; they were its essence' (Alan Bullock: *Hitler: A Study in Tyranny*, The Companion Book Club, London, 1954, p327).

## Summary

especially the Jews in 'The Final Solution'! All this flowed inexorably and openly – brazenly – from 'the survival of the fittest'. There is no baulking the fact. Take, as a couple of examples, Hitler's *Mein Kampf* and the 1935 Nazi film 'The Inheritance'.

Incidentally – and this is very important – the Nazi scheme was not the responsibility of a tiny number at the top – a few perverted, all-powerful tyrants. That is a myth! Historical, academic, research on a massive number of files, which the Nazis failed to destroy, has proved beyond doubt that while it was a mere handful of Gestapo in a town which effectively ran the procedure, those with the power of life and death in their hands were not working alone, imposing their sadism on the population: the secret police depended on the heaps – and I mean heaps – of letters they received from 'ordinary' locals, letters which gave them 'information' about people the informants thought 'unfit' (for whatever reason). The Gestapo simply did the paperwork involved in the process of having these 'misfits' eliminated.[3]

Evolutionists may try to argue that this was an aberration, but they are whistling in the wind. Yes, Hitler was remorseless in his logic, but his rationale was 'the survival of

---

[3] To explore this further would take me beyond the remit of this booklet, but the fact is, the higher echelon of the Nazi regime issued orders to its subordinates to encourage – or, at the very least, turn a blind eye to – lawless reprisals which broke out among disparate groups in the occupied territories. Countless atrocities were committed as a result. It was not just the Nazis. Just one statistic must suffice. In 1939, seventeen days after Germany invaded Poland from the West, Russia invaded from the East. Nikita 'Khrushchev ruthlessly suppressed any sections of the population who might oppose Soviet power: priests, officers, noblemen, intellectuals were kidnapped, murdered and deported to eliminate the very existence of Poland. By November 1940, one tenth of the population, or 1.17 million innocents, had been deported. Thirty percent of them were dead by 1941; 60,000 were arrested and 50,000 shot' (Simon Sebag Monetefiore: *Stalin: 1939-1953*, Phoenix, London, 2004, p15).

## Summary

the fittest'. The Nazis took evolution seriously: they gloried in the survival of the fittest; they wallowed in it.[4] After all, they were only following evolution's fundamental principle, and following it to the letter: violent competition is the order of the day; perpetual struggle should be encouraged, not apologised for; survival and reproduction of the fittest are the vital drivers, and nothing must stand in their way.

---

[4] Hitler: 'The idea of struggle is as old as life itself, for life is only preserved because other living things perish through struggle... In this struggle, the stronger, the more able, win, while the less able, the weak, lose. Struggle is the father of all things... It is not by the principles of humanity that man lives or is able to preserve himself above the animal world, but solely by means of the most brutal struggle... If you do not fight for life, then life will never be won' (Hitler's speech at Kulmbach, 5th Feb. 1928, quoted by Bullock p31).

# Three Questions Asked and Answered

1. Why do we still have a sense of fairness, right and wrong? If evolution is right, mankind should have no sense of pity,[1] no sense of wrongdoing, no sense of accountability, no conscience. And yet all of us do have a conscience. Evolutionists are no exception; they, too, have a conscience, a conscience which never gives up. Why? Where does it come from?

God has instilled it in us; it is written deep within each and every one of us. And we cannot shake it off. Come what may, stifle it as much as we can, sear it for all we are worth, devise any philosophy we like to rid ourselves of it, conscience is always there – dogging our every footstep, nagging at us, reminding us. Experience proves it. Mark Twain observed it: 'Man is the only created being[2] that blushes. Or needs to'.[3] I would add, or can do! How does evolution account for this gulf between men and the rest of creation?

As I say, God has instilled this conscience in us men; it is written deep within each and every one of us. Scripture declares it:

> [Men] show that the work of the law is written on their hearts, while their conscience also bears witness, and their conflicting thoughts accuse or even excuse them (Rom. 2:15).

Indeed, let me quote the entire paragraph:

---

[1] Nietzsche tried to persuade mankind that pity was wrong. Why did he need to do this? Why does mankind have and show pity?
[2] Twain had 'animal'.
[3] Mark Twain: *Following the Equator: A Journey Around the World*.

> For all who have sinned without the law will also perish without the law, and all who have sinned under the law will be judged by the law. For it is not the hearers of the law who are righteous before God, but the doers of the law who will be justified. For when Gentiles, who do not have the law, by nature do what the law requires, they are a law to themselves, even though they do not have the law. They show that the work of the law is written on their hearts, while their conscience also bears witness, and their conflicting thoughts accuse or even excuse them on that day when, according to my gospel, God judges the secrets of men by Christ Jesus (Rom. 2:12-16).

I say again, despite Darwin and his disciples, nothing can take away this inbuilt sense of right and wrong, this sense of fitness, this sense of accountability in man. Witness the excuses men make for themselves and their behaviour! Witness the obvious pressure man feels to justify himself before others – and within himself! The theory of evolution will never eliminate conscience from men and women. Nor can it satisfactorily account for it. The Bible, and the Bible only, tells us why we have a conscience.

2. But how does the Bible explain away the obvious struggle in nature, the way in which nature is perpetually engaged in a battle to survive, even to conquer? Take influenza. When a flu virus mutates – as it does frequently – and becomes lethal to humans, for a while it wreaks havoc upon its host, mankind. But it comes face to face with the law of diminishing returns. Take the 1918 pandemic which killed 50 million people worldwide. After a while, in addition to killing so many of the host it had infected (thus destroying itself in the process), the virus found that other humans became increasingly resistant. Even so, the virus responded by mutating yet again, and thus renewed its attack. The struggle for survival was a to-the-death battle between the virus and its victims. And that is just one example of the constant struggle in nature. In the words of Alfred Lord

## Three Questions Asked and Answered

Tennyson: 'Nature is red in tooth and claw'.[4] How does the Bible explain this away?

It doesn't! It does not explain it *away*. It explains it. As it makes clear, nature – including man – is in bondage, corrupted, engaged in a life-and-death struggle – a losing struggle at that.[5] This, however, is not because of evolution and 'the survival of the fittest' in its drive to produce new and 'fitter' species, but because of the curse imposed by God as a result of Adam's fall (Gen. 3:14-19; Rom. 5:12-21; 8:19-22; 1 Cor. 15:22,42-49). This, not evolution, accounts for the struggle evident in nature.

3. Admitting the grim consequences of evolution, even so hasn't Christianity got its own consequences? Evolution has to bear its responsibility – what about Christianity? Doesn't Christianity have its own can to carry? Look at the bloodshed, misery and torment caused in the world in the name of religion – Christianity, not excepted.

I concede the point. I go further: I echo and proclaim it. Religion has wrought untold harm in the world. And I deplore it. Religion – all religion, including Darwinism – is an abomination.

---

[4] Alfred Lord Tennyson's 'In Memoriam...', 1850: 'Who trusted God was love indeed/ And love creation's final law/ Though nature, red in tooth and claw/ With ravine, shriek'd against his creed'. Tennyson was posing questions about the apparent conflict between love as the basis of the Christian religion and the callousness of nature. The poem played (and continues to play) its part in the debate over Charles Darwin's theories on natural selection, as expressed in *The Origin of Species*, 1859. Richard Dawkins used 'red in tooth and claw' in *The Selfish Gene*, to summarise the behaviour of all living things based on 'the survival of the fittest'.

[5] Whatever advances in medicine, death is still inevitable. As one disease is conquered, another, more virulent, disease springs up. We are already paying the price of our cavalier over-use of antibiotics. Pandemics are confidently predicted. We are destroying the environment. And so on.

But we need to be clear. In the case of the religion known as Christianity, we are not talking about Christ and his new covenant, the gospel. Rather, we are speaking of Christendom – by which I mean the so-called 'Christian world', 'cultural Christianity', 'Churchianity', 'religion in the name of Christ'. Christendom is a monstrosity. And Christendom truly is responsible for a massive – horrendous – catalogue of bloodshed, abuse, slavery and terror. So I say again, I admit it, I deplore it. But, as several of my books make clear, Christendom is the work and tool of Satan, and is an utter aberration of the glories of the new covenant established by Christ. It may use much of Christ's language, but it warps the new covenant beyond recognition.

And this is the material point. Christendom has nothing in common with Christ's new covenant; indeed, its appalling record comes about as a result of its rejection of Christ's covenant! In contrast to this, however, the horrors of evolution are a direct and inevitable consequence of evolution's very own principles. In short, you cannot be an evolutionist without accepting and glorying in the doctrine of 'the survival of the fittest'.

Let evolutionists try to find fault with the Lord Jesus Christ and his new covenant! What would the world be like if all men lived according to the law of Christ?

True believers in the Lord Jesus Christ deplore Christendom and all it stands for;[6] they proclaim and seek to live by the gospel of the Lord Jesus Christ as recorded in the Scriptures. True believers fully accept – indeed, they glory in – all the consequences of Christ's gospel – not least, salvation from sin, and eternal bliss to come in everlasting glory, for all who are in Christ. Evolutionists must do the same with their belief system. Let them preach 'the survival of the fittest'.

---

[6] At least, they ought to! Alas, Christendom is so ingrained, many believers just do not realise how grievously this invention has marred the new covenant: they often think that Christendom is the biblical norm, when it is anything but.

Let them glory in it. Let them stop trying to get round its consequences so as to make them palatable to the God-given conscience in men.

# *A Positive Note to End On*

I do not want to leave this booklet there. Surely there must be hope. Are we really doomed to 'the survival of the fittest'? Is that the best we have? Most of us are anything but 'the fittest'. Indeed, we are all sinners. Is there no hope? Is Woody Allen right after all? Remember what he said. Listen to the hopelessness in his voice:

> More than [at] any other time in history, mankind faces a crossroads. One path leads to despair and utter hopelessness. The other, to total extinction. Let us pray we have the wisdom to choose correctly.[1]

From where I am standing, it seems that most end up adopting the mantra the sacred Teacher observed men repeating in his day:

> Surely the fate of human beings is like that of the animals; the same fate awaits them both: As one dies, so dies the other. All have the same breath; humans have no advantage over animals.

Adopting such a stance, there can be only one conclusion:

> Everything is meaningless [or vain; that is, passing, temporary, insubstantial, failing to satisfy] (Eccles. 3:19).

Vain? Ephemeral? Is it enough to know that my life has been spent merely producing descendants 'fitter' for their remorseless chase around the treadmill, and for them, in turn, to produce descendants who will be 'fitter' still? Is that all there is to it? Is this the sum and substance of life? For the evolutionist, yes! What else can it be?

But for the Christian, things are very different:

> When I consider your [that is, God's] heavens, the work of your fingers, the moon and the stars, which you have set in

---

[1] Woody Allen: 'My Speech to the Graduates', *The New York Times*, August 10th, 1979.

## A Positive Note to End On

place, what is mankind that you are mindful of them – human beings – that you care for them? You have made them a little lower than the angels and crowned them with glory and honour. You made them rulers over the works of your hands; you put everything under their feet: all flocks and herds, and the animals of the wild, the birds in the sky, and the fish in the sea, all that swim the paths of the seas. LORD, our LORD, how majestic is your name in all the earth! (Ps. 8:3-9).

And believers have this confident expectation:

Our present sufferings are not worth comparing with the glory that will be revealed in us. For the creation waits in eager expectation for the children of God to be revealed. For the creation was subjected to frustration, not by its own choice, but by the will of the one who subjected it, in hope [that is, in confident expectation] that the creation itself will be liberated from its bondage to decay and brought into the freedom and glory of the children of God. We know that the whole creation has been groaning as in the pains of childbirth right up to the present time. Not only so, but we ourselves, who have the firstfruits of the Spirit, groan inwardly as we wait eagerly for our adoption to sonship, the redemption of our bodies. For in this hope we were saved (Rom. 8:18-24).

How did Paul arrive at such a place? Read the preceding chapters of Romans, and all will become clear. We are all sinners, all under the wrath and condemnation of God. We cannot save ourselves. But God has acted in history – sending his Son to live and die for sinners, to atone for their sins, and sinners trusting Christ are at once fully forgiven and justified in the sight of God: the blood of Christ washes them from sin, and the righteousness of Christ clothes them. And that leads Paul to set out the glorious hope in the passage I just quoted. And there is more! The apostle moves straight on to declare that nothing shall ever take away this hope for the believer:

What, then, shall we say in response to these things? If God is for us, who can be against us? He who did not spare his own Son, but gave him up for us all – how will he not also,

## A Positive Note to End On

along with him, graciously give us all things? Who will bring any charge against those whom God has chosen? It is God who justifies. Who then is the one who condemns? No one. Christ Jesus who died – more than that, who was raised to life – is at the right hand of God and is also interceding for us. Who shall separate us from the love of Christ? Shall trouble or hardship or persecution or famine or nakedness or danger or sword?... No, in all these things we are more than conquerors through him who loved us. For I am convinced that neither death nor life, neither angels nor demons, neither the present nor the future, nor any powers, neither height nor depth, nor anything else in all creation, will be able to separate us from the love of God that is in Christ Jesus our Lord (Rom. 8:31-39).

And that brings me to why I have written this booklet – not to score points in a debate with evolutionists, but to point helpless and hopeless sinners – and Darwinism leaves sinners utterly helpless and hopeless – to the only hope, the glorious hope found in the Lord Jesus Christ alone.

More than that: I appeal to you to trust the Saviour and his finished work. I assure you that everyone who calls on the name of the Lord is truly saved (Rom. 10:13). So trust him now!

www.ingramcontent.com/pod-product-compliance
Lightning Source LLC
Chambersburg PA
CBHW030118230526
45469CB00005B/1693